Elementary Calculus

Elementary Calculus

George N. Frempong

Contents

Preface

This book has been developed base on my experience and insights received from colleagues and students. My goal is to present content that is easy to understand. The text is written in a manner that will help students both learn and retain concepts. I have provided careful explanations of concepts along with examples that are easy to follow. The clarity of the text provides students with more opportunities to learn, practice, and apply what they have learned.

1 Limits of Functions

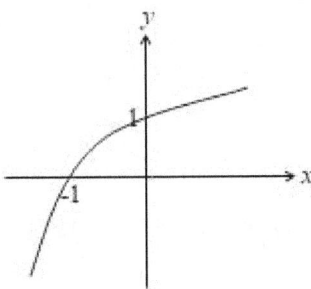

Limits are used to describe the behavior of a function as the values of the function approach or become closer and closer to some particular number. If a limit exists, it is unique.

Consider the graph of the function $f(x)$ as shown in the figure above. If we take x very close to zero through positive values, the value of the function approaches 1, This value is called the right-hand limit of the function. Also, the value of the function approaches 1 as we take x very close to zero through negative values. This value is called left-hand limit of the function. When the left-hand and right-hand limits are identical, their common value is called the limit of the function. For the above example, the limit of the function as x approaches 0 is 1, written

$$\lim_{x \to 0} f(x) = 1$$

and read "the limit as x approaches 0 of the function f is 1.

The definition of limit describe what happens to a function $f(x)$ when x is near a number, say a. Note that, the limit of a function as we approach a point is not necessarily the same as the value of the function at that point.

If the value of function $f(x)$ approaches l as x takes values closer and closer to a on both side, then l is the limit of $f(x)$ as x approaches a, written

$$\lim_{x \to a} f(x) = l$$

The following properties of limit will be extremely useful in computing the limits of functions.

2 Elementary Calculus

1. $\lim_{x \to a} c = c$ where c is a constant.

If $f(x)$ and $g(x)$ are functions such that

$$\lim_{x \to a} f(x) = l \ \text{ and } \ \lim_{x \to a} g(x) = m, \text{then}$$

2. $\lim_{x \to a} (f(x) \pm g(x)) = \lim_{x \to a} f(x) \pm \lim_{x \to a} g(x) = l \pm m$

3. $\lim_{x \to a} (f(x) \cdot g(x)) = \lim_{x \to a} f(x) \cdot \lim_{x \to a} g(x) = l \cdot m$

4. $\lim_{x \to a} \left(\dfrac{f(x)}{g(x)} \right) = \dfrac{\lim_{x \to a} f(x)}{\lim_{x \to a} g(x)} = \dfrac{l}{m}, provided \ that \ m \neq 0$

Examples

Evaluate

1. $\lim_{x \to 2} (2x - 1)$

$$\lim_{x \to 2} (2x - 1) = \lim_{x \to 2} 2x - \lim_{x \to 2} 1$$

$$= 2 \lim_{x \to 2} x - 1$$

$$= 2(2) - 1$$

$$= 3$$

2. $\lim_{x \to -2} (x^2 + 3x + 2) = \lim_{x \to -2} x^2 + 3 \lim_{x \to -2} x + \lim_{x \to -2} 2$

$$= (-2)^2 + 3(-2) + 2$$

$$= 0$$

3. $\lim_{x \to 4} \dfrac{3x + 2}{x - 2} = \dfrac{3 \lim_{x \to 4} x + \lim_{x \to 4} 2}{\lim_{x \to 4} x - \lim_{x \to 4} 2}$

$$= \dfrac{3(4)+2}{4-2}$$

$$= 7$$

4. $\lim\limits_{x \to 1} \dfrac{x^2 - 1}{x - 1}$

The denominator approaches 0 as x approaches 1. Because division by 0 is undefined, you cannot substitute 1 for x. Begin by factorizing the numerator.

$$\lim_{x \to 1} \frac{x^2 - 1}{x - 1} = \lim_{x \to 1} \frac{(x - 1)(x + 1)}{x - 1}$$

$$= \lim_{x \to 1} (x + 1)$$

$$= \lim_{x \to 1} x + \lim_{x \to 1} 1$$

$$= 1 + 1$$

$$= 2$$

5. $\lim\limits_{x \to \infty} \dfrac{6x^2 + 3x + 2}{4x^2 - x - 7}$

We divide the numerator and denominator by the highest power of x present in the denominator and use the result

$$\lim_{x \to \infty} \frac{1}{x} = 0$$

In this case, the highest power of x is x^2.

$$\lim_{x \to \infty} \frac{6x^2 + 3x + 2}{4x^2 - x - 7} = \lim_{x \to \infty} \frac{6 + \dfrac{3}{x} + \dfrac{2}{x^2}}{4 - \dfrac{1}{x} - \dfrac{7}{x^2}}$$

$$= \frac{3}{2}$$

Exercise 1

Find the indicated limit:

1. $\lim\limits_{x \to 2} (2x + 5)$ 2. $\lim\limits_{x \to 3} (x + 3)$

3. $\lim\limits_{x \to 0} (x^2 + 4)$ 4. $\lim\limits_{x \to 4} (7 - 3x)$

5. $\lim\limits_{x \to -1} 3x^2$

6. $\lim\limits_{x \to 3}(7 - 3x)$

7. $\lim\limits_{x \to -3} (x^2 - 6x - 15)$

8. $\lim\limits_{x \to -2} (x^2 - x - 2)$

9. $\lim\limits_{x \to 4} \sqrt{2x^2 - 7}$

10. $\lim\limits_{x \to 2} \sqrt[3]{5x^2 + 7}$

11. $\lim\limits_{x \to 2} \dfrac{2x - 1}{x + 6}$

12. $\lim\limits_{x \to 3} \dfrac{x^2 - 9}{x - 3}$

13. $\lim\limits_{x \to -2} \dfrac{x^2 + x - 2}{x + 2}$

14. $\lim\limits_{x \to 1} \dfrac{x^2 + x - 2}{x + 2}$

15. $\lim\limits_{x \to \infty} \dfrac{2x^2 + 3x - 4}{1 - x^2}$

16. $\lim\limits_{x \to \infty} \dfrac{5x^2 + 4x + 3}{5x^2 - x - 7}$

In Exercises 17 – 20, calculate

$$\lim\limits_{h \to 0} \dfrac{f(a + h) - f(a)}{h}$$

17. $f(x) = 2x$

18. $f(x) = 3x^2$

19. $f(x) = 2x + 3$

20. $f(x) = 2x^2 - 3$

2 The Derivative

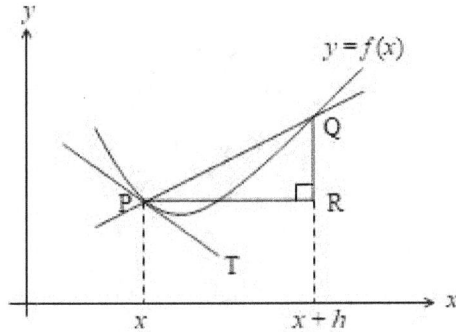

The figure shows the graph of a function $y = f(x)$. Let P and Q be two points with x-coordinates at x and $x + h$ respectively.

The slope of the chord PQ is

$$\frac{QR}{PR} = \frac{f(x + h) - f(x)}{h}$$

As the point Q moves closer and closer to the point P, h approaches zero. When Q coincides with P the chord PQ becomes the tangent PT to the curve at P. Thus, the slope of the tangent at P is the limit of the slopes of the corresponding chords as h approaches 0. The tangent have slope given by

$$\lim_{h \to 0} \frac{f(x + h) - f(x)}{h}$$

This limit defines another function called the derivative of f with respect to x. The derivative of $y = f(x)$ may be written in any of the following ways

$$f'(x) \qquad \frac{dy}{dx} \qquad \frac{d}{dx} f(x) \quad \text{or} \quad D_x f(x)$$

The process of computing the derivative is called differentiation.

Note that, the slope of the tangent is equal to the derivative. The slope of the tangent at a point is also called the slope of the curve at that point.

Consider the function $f(x) = x^2$.

The slope of the tangent to the curve at $(x, f(x))$ is given by

$$\lim_{x \to 0} \frac{f(x+h) - f(x)}{h} = \lim_{x \to 0} \frac{(x+h)^2 - x^2}{h}$$

$$= \lim_{x \to 0} \frac{2xh + h^2}{h}$$

$$= \lim_{h \to 0} (2x + h)$$

$$= 2x$$

The function $f'(x) = 2x$ is the derivative of $f(x) = x^2$.

Differentiation from first principles

The method of using the definition to calculate the derivative of a function is called differentiation from first principles.

Example

Find the derivative of the following functions from first principles.

1. $f(x) = x^3$

$$f'(x) = \lim_{h \to 0} \frac{f(x+h) - f(x)}{h}$$

$$= \lim_{h \to 0} \frac{(x+h)^3 - x^3}{h}$$

$$= \lim_{h \to 0} \frac{(x^3 + 3x^2h + 3xh^2 + h^3) - x^3}{h}$$

$$= \lim_{h \to 0} \frac{3x^2h + 3xh^2 + h^3}{h}$$

$$= \lim_{h \to 0} (3x^2 + 3xh + h^2)$$

$$= 3x^2$$

2. $f(x) = 2z$

$$f'(x) = \lim_{h \to 0} \frac{f(x+h) - f(x)}{h}$$

$$= \lim_{h \to 0} \frac{2(x+h) - 2x}{h}$$

$$= \lim_{h \to 0} \frac{2x + 2h - 2x}{h}$$

$$= \lim_{h \to 0} \frac{2h}{h}$$

$$= \lim_{h \to 0} 2$$

$$= 2$$

3. $f(x) = -3$

$$f'(x) = \lim_{h \to 0} \frac{f(x+h) - f(x)}{h}$$

$$= \lim_{h \to 0} \frac{-3 - (-3)}{h}$$

$$= \lim_{h \to 0} \frac{0}{h}$$

$$= 0$$

Exercise 2(a)

Find the derivative of each of the following functions from first principles:

1. $f(x) = x$ 2. $f(x) = 5x$

3. $f(x) = 3x^2$ 4. $f(x) = -x^2$

5. $f(x) = 2$ 6. $f(x) = -\frac{1}{2}x^2$

7. $f(x) = \frac{2}{3}x^3$ 8. $f(x) = -3x^2$

9. $f(x) = 2x^3$

10. $f(x) = 3x^2 - 1$

11. $f(x) = 2 - 3x^2$

12. $f(x) = x^2 + 3$

General Rule for Calculating Derivatives

In the preceding section we used a method that shows how to use the definition to find the derivative of a function. The results of the examples suggest the following rules for computing derivatives. These rules make the calculation of derivative much easier.

Power Rule

If $f(x) = x^n$ for any real number $n > 0$, then $f'(x) = n x^n$

Notice that the derivative of $f(x) = x^n$ is found by multiplying the function by the exponent n and decreasing the exponent on x by 1.

Example

Find the derivative of each function.

(a) $f(x) = x^5$

$$f'(x) = 5 \cdot x^{5-1} = 5x^4$$

(b) $y = x^{-3}$

$$\frac{dy}{dx} = -3x^{-3-1} = -3x^{-4}$$

(c) $f(x) = x^{3/2}$

$$f'(x) = \frac{3}{2}x^{\frac{3}{2}-1} = \frac{3}{2}x^{\frac{1}{2}}$$

The results can be obtained mentally.

Exercise 2(b)

Find the derivative of each function.

1. $f(x) = -x$

2. $f(x) = 8$

3. $f(x) = x^4$

4. $f(x) = x^6$

5. $f(x) = -x^{-3}$ 6. $f(x) = x^{-1}$

7. $f(x) = x^{-8}$ 8. $f(x) = x^{-3/4}$

9. $f(x) = -x^{-1/2}$ 10. $f(x) = x^{-3/2}$

11. $f(x) = -x^{-2/3}$ 12. $f(x) = -x^{1/3}$

Derivative of a Constant Times a Function

If f and g are differentiable functions, and c a constant, then

$$\frac{d}{dx}[cf(x)] = c\frac{d}{dx}f(x)$$

To differentiate leave the constant and differentiate the function and then multiply the result by the constant.

Example

Find the derivative of the following:

(a) $f(x) = -3x^2$

$$f'(x) = \frac{d}{dx}(-3x^2)$$

$$= -3 \times 2x$$

$$= -6x$$

(b) $y = \frac{1}{2}x^6$

$$\frac{dy}{dx} = \frac{1}{2} \times 6x^5$$

$$= 3x^5$$

Exercise 2(c)

Find the derivative of each of the following functions:

1. $f(x) = -2x$ 2. $f(x) = 3x^4$

3. $f(x) = -2x^5$ 4. $f(x) = \frac{1}{4}x^8$

5. $f(x) = \frac{2}{3}x^6$ 6. $f(x) = \frac{2}{5}x^{-5}$

7. $f(x) = 2x^{-3}$ 8. $f(x) = -4x^{-3}$

9. $f(x) = 3x^{-2}$ 10. $f(x) = 3x^{1/3}$

11. $f(x) = -2x^{3/2}$ 12. $f(x) = 6x^{1/2}$

13. $f(x) = -6x^{-2/5}$ 14. $f(x) = 8x^{-3/4}$

15. $f(x) = 9x^{-2/3}$ 16. $f(x) = \frac{2}{3}x^{-6}$

Sum or Difference Rule

If f and g are differentiable functions, then

$$\frac{d}{dx}[f(x) \pm g(x)] = \frac{d}{dx}f(x) \pm \frac{d}{dx}g(x)$$

To differentiate a sum or a difference of functions differentiate term by term

Example

Find the derivative of $f(x) = x^3 + 3x^2 - 2x$

$$f'(x) = 3x^2 + 6x - 2$$

Exercise 2(d)

Differentiate each of the following functions with respect to the variable.

1. $f(x) = 2x^3 + 3x^4$ 2. $f(x) = 3x^5 - 2x$

3. $f(t) = 3 - t^4 + t^5$ 4. $f(t) = 7 - 2t^{-1}$

5. $f(t) = 4t^3 + 3t$ 6. $f(x) = 5x^4 - 3x^5$

7. $f(x) = 5 - x^{-3}$ 8. $f(x) = 2x^3 - 3x^{-2}$

9. $f(t) = 3t^4 - 4t + 2$ 10. $f(t) = 2t^2 - 3t$

11. $f(x) = 6x^2 + 5x + 2$ 12. $f(t) = \frac{1}{3}t^3 - \frac{1}{2}t^{-2}$

13. $f(x) = 3x^{1/3} - 2x^{-1/2}$ 14. $f(t) = 4t^{-3/4} + \frac{1}{3}t^3$

15. $f(x) = 2 - 6x^{-2/3}$ 16. $f(x) = 1 + \frac{1}{x^3}$

17. $f(x) = 3x - \frac{1}{\sqrt{x}} + \frac{1}{x}$ 18. $f(x) = \frac{5}{x^2} - \frac{1}{\sqrt{x^3}} + 2$

19. $f(t) = t^2 + \frac{4}{t^3}$ 20. $f(x) = 2x(3 - x)$

21. $f(x) = 3(x - 2)^2$ 22. $f(x) = \frac{3x^2 - 2x^3}{3x}$

23. $f(x) = \frac{3x^2 + 2x - 1}{x}$ 24. $f(t) = \frac{(t+3)(2t-1)}{t^2}$

Derivatives of Products and Quotients

Product Rule

If f and g are differentiable functions, then

$$\frac{d}{dx}[f(x) \cdot g(x)] = f(x)\frac{d}{dx}g(x) + g(x)\frac{d}{dx}f(x)$$

To differentiate a product of two functions, leave the first function and differentiate the second function + leave the second function and differentiate the first function.

Example

Find the derivative of the following:

(a) $f(x) = x^3(2x - 3)$

$$f'(x) = x^3 \times 2 + (2x - 3) \times 3x^2$$

$$= 2x^3 + 6x^3 - 9x^2$$

$$= 8x^3 - 9x^2$$

The same result is obtained if you expand the bracket first and then find the derivative.

$$f(x) = 2x^4 - 3x^3$$

$$f'(x) = 8x^3 - 9x^2$$

Often it will be quite difficult or unnecessary to multiple out in order to differentiate term by term.

(b) $f(x) = (2x + 3)(x^3 - 4)$

$$f(x) = (2x + 3)(x^3 - 4)$$

$$f'(x) = (2x + 3) \cdot 3x^2 + (x^3 - 4) \cdot 2$$

$$= 6x^3 + 9x^2 + 2x^3 - 8$$

$$= 8x^3 + 9x^2 - 8$$

Exercise 2(e)

Find the derivative of each function.

1. $f(x) = x^3(4x^2 - 3)$

2. $f(x) = (2x - 3)(x + 4)$

3. $f(x) = 3x^2(2x + 1)$

4. $f(x) = (2x + 1)(3x - 2)$

5. $f(x) = x(2x^3 - 3x^2)$

6. $f(x) = x(3 - 2x)(2 + x)$

7. $f(x) = \sqrt{x}(2x + 3)$

8. $f(x) = (1 - x^2)(1 + 2x^2)$

9. $f(x) = \sqrt[3]{x}(2 + x)$

10. $f(x) = 3x^{-2}(2x - 3)$

11. $f(x) = (x^4 - 1)(x^3 + 1)$

12. $f(x) = (2x - 5)(3x^2 + 2)$

Quotient Rule

If f and g are differentiable functions and $g(x) \neq 0$, then

$$\frac{d}{dx}\left[\frac{f(x)}{g(x)}\right] = \frac{g(x)\frac{d}{dx}f(x) - f(x)\frac{d}{dx}g(x)}{[g(x)]^2}$$

To differentiate a quotient of two functions, leave the denominator and differentiate the numerator – leave the numerator and differentiate the denominator and divide all by the square of the denominator.

Example

Find the derivative of the following:

(a) $y = \dfrac{2x}{3x-1}$

$y = \dfrac{2x}{3x-1}$

$\dfrac{dy}{dx} = \dfrac{(3x-1)\frac{d}{dx}(2x)-2x\frac{d}{dx}(3x-1)}{(3x-1)^2}$

$= \dfrac{(3x-1)\times 2 - 2x\times 3}{(3x-1)^2}$

$= \dfrac{6x-2-6x}{(3x-1)^2}$

$= -\dfrac{2}{(3x-1)^2}$

Exercise 2(f)

Find the derivative of each function.

1. $f(x) = \dfrac{x}{x+1}$

2. $f(x) = \dfrac{x+1}{x-1}$

3. $f(x) = \dfrac{1}{x^2-1}$

4. $f(x) = \dfrac{2}{3x^2-1}$

5. $f(x) = \dfrac{3x}{x^2-2}$

6. $f(x) = \dfrac{3x^2-5}{2x-7}$

7. $f(x) = \dfrac{1-\sqrt{x}}{1+\sqrt{x}}$

8. $f(x) = \dfrac{1+x^2}{1-x^2}$

9. $f(x) = \dfrac{x^2}{x+1}$

10. $f(x) = \dfrac{x^2}{x^2-1}$

11. $f(x) = \dfrac{x^2}{2x+3}$

12. $f(x) = \dfrac{\sqrt{x}}{\sqrt{x}-1}$

The Chain Rule

Many functions are composite functions of simpler functions. For instance, if f, g and h are functions such that $f(x) = x^3$, $g(x) = 3x + 2$ and $h(x) = (3x + 2)^3$, then $h(x) = f[g(x)]$. If $y = f(x)$ and $u = g(x)$, then $y = f[g(x)]$ and

$$\frac{dy}{dx} = \frac{dy}{du} \cdot \frac{du}{dx}$$

The chain rule is used to find derivatives of a composite function.

Example

Find the derivative of each function.

(a) $f(x) = (2x - 3)^4$

Let $y = (2x - 3)^4$,

Then $y = u^4$ and $u = 2x - 3$. So,

$$\frac{dy}{du} = 4u^3$$

and $\frac{du}{dx} = 2$

By the chain rule

$$\frac{dy}{dx} = \frac{dy}{du} \cdot \frac{du}{dx}$$

$$= 4u^3 \times 2$$

$$= 8u^3$$

Replacing u with $2x - 3$ gives

$$\frac{dy}{dx} = 8(2x - 3)^3$$

(b) $y = (2x + 1)^{-3}$

Let $u = 2x + 1$, so $y = u^{-3}$

$$\frac{dy}{du} = -3u^{-4}$$

and $\frac{du}{dx} = 2$

Then $\frac{dy}{dx} = \frac{dy}{du} \cdot \frac{du}{dx}$

$$= -3u^{-4} \cdot 2$$

$$= -6u^{-4}$$

Replacing u with $2x + 1$ gives

$$\frac{dy}{dx} = -6(2x + 1)^{-4}$$

$$\frac{dy}{dx} = -6(2x + 1)^{-4}$$

Exercise 2(g)

Find the derivative of each function.

1. $f(x) = (x - 5)^3$ 2. $f(x) = (3x - 2)^5$

3. $f(x) = (1 - 2x^2)^6$ 4. $f(x) = (x^2 + 1)^4$

5. $f(x) = (x^3 - 2)^7$ 6. $f(x) = (2x + 3)^{-3}$

7. $f(x) = (3 - 2x)^{-4}$ 8. $f(x) = (4x + 3)^{-2}$

9. $f(x) = (1 - 2x^2)^{-1}$ 10. $f(x) = (x^2 - 1)^{1/2}$

11. $f(x) = (3x + 2)^{-2/3}$ 12. $f(x) = (2x^2 - 3)^{3/4}$

The following is an alternative version of the Chain Rule.

If $y = f[g(x)]$, then

$$\frac{dy}{dx} = f'[g(x)] \cdot g'(x)$$

That is, to find the derivative of $f[g(x)]$, find the derivative of the outer function, then multiply by the derivative of the inner function.

Example

Find the derivative of each function.

(a) $f(x) = (3 - 2x)^5$

$$f'(x) = 5(3 - 2x)^4 \cdot -2$$

$$= -10(3 - 2x)^4$$

(b) $f(x) = (x^3 + 2)^{1/3}$

$$f'(x) = \frac{1}{3}(x^3 + 2)^{-2/3} \cdot 3x^2$$

$$= x^2(x^3 + 2)^{-2/3}$$

(c) $f(x) = (x^2 + 3x)^{-3}$

$$f'(x) = -3(x^2 + 3x)^{-4} \cdot (2x + 3)$$

$$= -3(2x + 3)(x^2 + 3x)^{-4}$$

Exercise 2(h)

Find the derivative of each function.

1. $f(x) = (x + 3)^4$

2. $f(x) = (3x - 2)^3$

3. $f(x) = (2x^2 + 3)^5$

4. $f(x) = (x^2 + 1)^{1/2}$

5. $f(x) = (2x^3 - 1)^{-1/3}$

6. $f(x) = (3 + x^2)^{-2}$

7. $f(x) = (1 - x^2)^3$

8. $f(x) = (3x^2 - 1)^{-3}$

9. $f(x) = (1 - x^2)^6$

10. $f(x) = \frac{1}{(x^2 - 3)^4}$

11. $f(x) = \frac{1}{\sqrt{2x^2 + 3}}$

12. $f(x) = \frac{1}{\sqrt[3]{x^3 + 3x}}$

3 Higher Order Derivatives

If a function f has a derivative f', then the derivative of f', if it exists, is the second derivative of f, written f''. The derivative of f'', if it exists, is called the third derivative of f. By continuing this process, you can find the fourth and other higher derivatives.

Second Order Derivative

The second order derivative of $y = f(x)$ can be written using any of the following notations.

$$f''(x) \qquad \frac{d^2y}{dx^2} \qquad \text{or} \qquad D_x^2[f(x)]$$

Example

Find the second derivative of each function.

(a) $f(x) = 3x^4 - 2x^3$

$$f'(x) = 12x^3 - 6x^2$$

$$f''(x) = 36x^2 - 12x$$

(b) $f(x) = (x^2 + 1)^3$

$$f'(x) = 3(x^2 + 1)^2 \cdot 2x$$

$$= 6x(x^2 + 1)^2$$

$$f''(x) = 6x \cdot 2(x^2 + 1) \cdot 2x + (x^2 + 1)^2 \cdot 6$$

$$= 24x^2(x^2 + 1) + 6(x^2 + 1)^2$$

$$= 6(x^2 + 1)(5x^2 + 1)$$

(c) $y = \dfrac{x}{1 + x}$

$$\frac{dy}{dx} = \frac{(1+x)\cdot 1 - x\cdot 1}{(1+x)^2}$$

$$= \frac{1+x-x}{(1+x)^2}$$

$$= \frac{1}{(1+x)^2}$$

$$\frac{d^2y}{dx^2} = \frac{(1+x)^2 \cdot 0 - 1 \cdot 2(1+x)}{(1+x)^4}$$

$$= \frac{-2(1+x)}{(1+x)^4}$$

$$= \frac{-2}{(1+x)^3}$$

Exercise 3

Find the second derivative of each function.

1. $f(x) = x^4 + x^2 - 2$

2. $f(x) = 3 - 2x + x^2 - x^3$

3. $f(x) = 2 + 3x^{-1}$

4. $f(x) = x^{-3} - 2x$

5. $f(x) = 3x + \frac{2}{x^2}$

6. $f(x) = 1 - \frac{4}{x^3}$

7. $f(x) = 1 - \sqrt[3]{x}$

8. $f(x) = \frac{1}{\sqrt[3]{x}} - \frac{2}{\sqrt{x}} + \frac{3}{x}$

9. $f(x) = (1 + x^3)^2$

10. $f(x) = (\sqrt{x} + 1)^2$

11. $f(x) = 2x(x-3)^3$

12. $f(x) = (x-2)(x+2)^2$

13. $f(x) = \frac{x}{x+1}$

14. $f(x) = \frac{1+x^2}{1-x^2}$

15. $f(x) = \frac{1}{\sqrt{1+x}}$

16. $f(x) = \frac{x^2}{\sqrt{x-1}}$

4 Applications of the Derivatives

Related Rates

Often, each quantity in an equation changes with time (or some other variable). For example, if a cylindrical container collects water from a tap, the volume, V, and depth, h of water increases with time t. Notice that the volume of water also increases as the depth of water increases. By the chain rule, we could obtain an equation which connects the rate of change of volume and depth. Since V is a function of h, and both V and h are functions of t, then by the chain rule

$$\frac{dV}{dt} = \frac{dV}{dh} \cdot \frac{dh}{dt}$$

Using this equation, we could calculate either the rate of change of volume or of depth, if the value of one of them is given.

Solving related rates problem usually involve three steps.

1. Find the functional relationship between the variables in the problem.

2. Using the chain rule write an equation connecting the rate of change of the variables.

3. Substitute for the given values and solve for the derivative giving the unknown rate.

Example

(a) The side of a cube is increasing at 5 cm s^{-1}, what is the rate of increase of the volume when $x = 3$ cm.

Let the length of the side of the cube be x cm. The volume, V of the cube and the length of the side x are related by

$$V = x^3$$

Find the derivative of V with respect to x.

$$\frac{dV}{dx} = 3x^2$$

Both V and x are functions of time t, so by the chain rule

$$\frac{dV}{dt} = \frac{dV}{dx} \cdot \frac{dx}{dt}$$

$$= 3x^2 \cdot 5$$

$$= 15x^2$$

Substituting 3 for x in this equation gives

$$\frac{dV}{dt} = 15 \cdot 3^2 = 135$$

The rate of increase of the volume is 135 cm^3 s^{-1}.

(b) Air is pumped into a spherical balloon at 8 cm^3 s^{-1}. How fast is the radius increasing when it is 2 cm?

The volume of the balloon is given by

$$V = \frac{4}{3}\pi r^3$$

Now, $\frac{dV}{dr} = 4\pi r^2$

By the chain rule

$$\frac{dV}{dt} = \frac{dV}{dt} \cdot \frac{dr}{dt}$$

$$8 = 4\pi r^2 \cdot \frac{dr}{dt}$$

$$\frac{dr}{dt} = \frac{1}{2\pi}$$

Thus, the radius is increasing at the rate of $1/2\pi$ cm s^{-1}.

Exercise 4(a)

1. The side of a square is increasing at 0.8 cm s^{-1}. At what rate is the area increasing at a time when the side is 10 cm?

2. The radius of a sphere is increasing at 0.2 cm s^{-1}. At what rate is the surface area increasing when the radius is 3 cm? At what rate is the volume increasing?

3. The area of a circle is increasing at 12 cm^2 s^{-1}. At what rate is the radius increasing, when it is 20 cm?

4. Water is poured into a cone of vertical angle 90° at 10 cm^3 s^{-1}. When the height of water is 15 cm, at what rate is it increasing?

5. The volume of a cube is decreasing at 6 cm^3 s^{-1}. When the side is 4 cm, what is the rate of decrease of (a) the side, and (b) the surface area?

6. A pump is inflating a spherical balloon. If the radius at a certain instant is 5 cm and it is increasing at a rate of 10 cm s^{-1}, at what rate is the pump working?

7. If air is pumped into a balloon at the rate of 0.50 m^3 s^{-1} at what rate will the radius be increasing when it is 5 m?

8. A spherical balloon is losing air at a rate of 18 m^3 s^{-1}. When its radius is 3 m, at what rate is it diminishing?

9. Liquid is dropping through a conical funnel at a rate of 5 cm^3 s^{-1}. When the depth of liquid in the funnel is x cm, its volume is $1/3\pi x^3$. Find the rate at which the level of liquid is falling when $x = 10$.

10. A funnel is made in the shape of a right circular cone of height 8 cm and base radius 6 cm. Water drains from the vertex of the funnel at the rate of 1.8 cm^3 s^{-1}. Find the rate at which the water level is dropping when the water has receded 4 cm from the top.

Small Changes

For any function $y = f(x)$, a small increment Δx in x causes a change, Δy in y. Thus, $\Delta y = f(x + \Delta x) - f(x)$

The rate of change of y with respect to x is given by

$$\frac{\Delta y}{\Delta x} = \frac{f(x + \Delta x) - f(x)}{\Delta x}$$

The derivative of the function f at x is defined as

$$\frac{dy}{dx} = \lim_{\Delta x \to 0} \frac{\Delta y}{\Delta x}$$

If the derivative exists, and Δx is a small nonzero real number, then

$$\frac{\Delta y}{\Delta x} \approx \frac{dy}{dx}$$

Multiplying both sides by Δx (provided $\Delta x \neq 0$) gives

$$\Delta y \approx \frac{dy}{dx} \cdot \Delta x$$

Example

(a) If the side of a cube is increased from 12 cm to 12.01 cm, what is the approximate increase in volume?

The volume of the cube is given by

$$V = x^3$$

Take the derivative of V with respective to x.

$$\frac{dV}{dx} = 3x^2$$

Now, $\Delta x = 12.01 - 12 = 0.01$

Using the small changes formula, with $\Delta x = 0.01$, we get

$$\Delta y \approx \frac{dV}{dx} \cdot \Delta x$$

$$= 3x^2(0.01)$$

$$= 0.03x^2$$

Substituting 12 for x gives

$$\Delta y = 0.03(12^2) = 4.32$$

The change in volume is approximately 4.32 cm³.

(b) Evaluate $\sqrt[3]{125.1}$

First, choose a number x such that it is close to 125.1 and has an exact cube root. Since, $y = \sqrt[3]{125} = 5$, we let $y = \sqrt[3]{x}$, where $x = 125$. So, $\Delta x = 0.1$

Rewrite $y = \sqrt[3]{x}$ as $y = x^{1/3}$. Then

$$\frac{dy}{dx} = \frac{1}{3}x^{-2/3} = \frac{1}{3\sqrt[3]{x^2}}$$

Using the small change formula, we have

$$\Delta y \approx \frac{dy}{dx} \cdot \Delta x$$

$$= \frac{1}{3\sqrt[3]{x^2}}\Delta x$$

Substituting 125 for x and 0.1 for Δx gives

$$\Delta y = \frac{1}{3\sqrt[3]{125^2}} \cdot 0.1 = 0.0013$$

Thus, $\sqrt[3]{125.1} = 5 + 0.0013 = 5.0013$

(c) The radius of a sphere is measured as 16 cm, with a possible error of 0.02 cm. What is the possible error in the volume?

The volume of a sphere is given by

$$V = \frac{4}{3}\pi r^3$$

$$\frac{dV}{dr} = 4\pi r^2$$

The error in volume is

$$\Delta V \approx \frac{dV}{dr}\Delta r$$

$$= 4\pi r^2 \Delta r$$

Substituting 16 for r and 0.02 for Δr gives

$$\Delta V \approx 4\pi \, (16^2)(0.02) = 64.3$$

The error in volume is approximately 64.3 cm^3.

Exercise 4(b)

1. Find the approximate increase in area of a square when its side changes from 10 cm to 10.1 cm.

2. A cube has side 8 cm. Find the approximate change in its volume if its side changes by 0.02 cm.

3. The radius of a sphere is measured as 6.5 cm, with possible error of 0.03 cm. What is the possible error in the volume?

4. The side of a square is measured with a possible error of 3 %, what is the approximate percentage error in the area?

5. The percentage error when measuring the area of a circle was 5 %. What was the approximate percentage error in its radius?

6. What is the error in area of a circle and its circumference if its radius is 0.2 % greater than it correct radius of 1 m?

7. An error of 1½ % is made in measuring the radius of a sphere. Find the percentage error in surface area.

8. An error of 2 % is made in measuring the radius of a sphere. What are the resulting errors in the calculation of its surface area and volume?

9. The height of a cylinder is 6 cm and its radius is 3 cm. Find the approximate increase in volume when the radius increases to 3.02 cm.

10. The volume of a sphere increases by 2 %. Find the corresponding percentage increase in surface area.

Use small changes to evaluate each of the following.

11. $\sqrt{53}$ 12. $\sqrt[3]{27.1}$ 13. $\sqrt[5]{32.01}$

14. $\sqrt{17.2}$ 15. $\sqrt[4]{81.02}$ 16. $\sqrt[3]{29}$

17. $\sqrt{147}$ 18. $\sqrt[5]{36}$ 19. $\sqrt[3]{1003}$

5 Local Maximum and Local Minimum

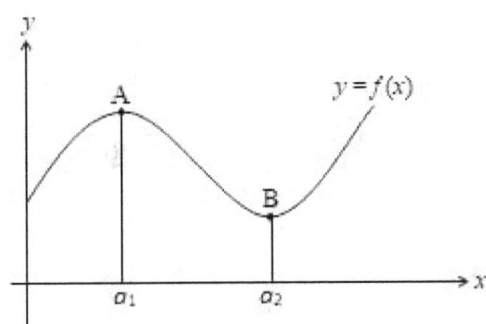

The figure above shows the graph of a function $y = f(x)$. Notice that the y-value of the function at A, $f(a_1)$ is greater than the y-values on either side of a_1. The number $f(a_1)$ is called the local maximum for f. The point $(a_1, f(a_1))$ is called the maximum point.

The y-value at B, $f(a_2)$ is less than the y-values on either side of a_2. The number $f(a_2)$ is called the local minimum for f, and the point $(a_2, f(a_2))$ is called the minimum point.

The points A and B are called turning points or stationary values.

The graph of a function f may have one or more turning points. If the graph of a function f has a turning point at $x = a$ then

1. $f(a)$ is a local maximum for f if $f(x) < f(a)$ for all x in some region around a.

2. $f(a)$ is a local minimum for f if $f(x) > f(a)$ for all x in some region around a.

When the function is given as an equation, we can use the derivative to determine these points.

Locating a Turning Point

A function has either a local maximum or a local minimum at $x = c$ if $f'(c) = 0$.

You can find the x-coordinate of a turning point by solving the equation $f'(x) = 0$. The corresponding y-coordinate can be found from the original equation.

To determine whether a turning point is a maximum or a minimum we consider how the slopes changes as they pass through the turning point.

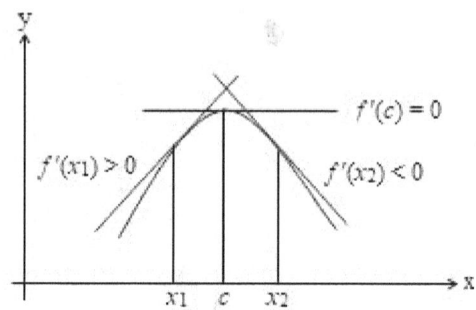

You can see from the figure above that at $x = c$ the slope is zero. The slope is positive on the left side of c and negative on the right side. The diagram shows that the turning point is a maximum.

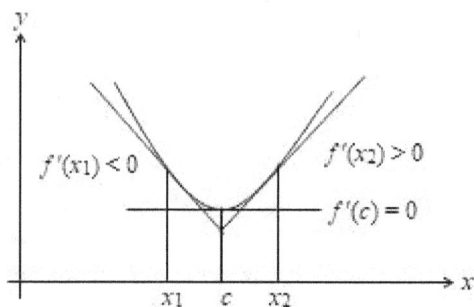

Observe that at $x = c$ the slope is zero. The slope is negative on the left side of c and positive on the right side. The diagram shows that the turning point is a minimum.

First Derivative Test

If $f'(c) = 0$, then

1. $f(c)$ is a local maximum of f if $f'(x)$ is positive on the left side of $x = c$ and negative on the right side.

2. $f(c)$ is a local minimum of f if $f'(x)$ is negative on the left side of $x = c$ and positive on the right side.

Example

1. Find all turning points and determine whether each turning point is a maximum or a minimum.

(a) $f(x) = x^2 - 6x - 15$

Begin by finding the derivative of f.

$$f'(x) = 2x - 6$$

Next set the derivative to 0 and solve for x.

$$2x - 6 = 0$$

$$x = 3$$

Thus, there is a turning point at $x = 3$.

To find the y-coordinate of the turning point substitute $x = 3$ into the original equation

$$f(3) = 3^2 - 6(3) - 15 = -24$$

The turning point is $(3, -24)$.

Finally, determine the signs of the slopes on either side of 3.

Choose a number close to 3 on the left side, and on the right side. Then substitute these numbers into $f'(x) = 2x - 6$.

Using 2.5 and 3.5 gives

$$f'(2.5) = 2(2.5) - 6 = -1 < 0$$

and $f'(3.5) = 2(3.5) - 6 = 1 > 0$.

Thus, the slope is negative on the left side of 3 and positive on the right side, indicating a minimum.

Thus, $(3, -24)$ is a minimum point.

(b) $f(x) = 3 + 2x - x^2$

The derivative is $f'(x) = 2 - 2x$

Set the derivative equal to 0.

$$2 - 2x = 0$$

giving $x = 1$

The y-coordinate of the turning point is

$$f(1) = 3 + 2(1) - 1 = 4$$

The turning point is (1, 4).

Now, determine the sign of $f'(x)$ on either side of 1.

Using 0.5 and 1.5 gives

$$f'(0.5) = 2 - 2(0..5) = 1 > 0$$

and $f'(1.5) = 2 - 2(1.5) = -1 < 0.$

Thus, the slope is positive on left side of 1 and negative on the right side, indicating a maximum.

Thus, (1, 4) is a maximum point.

2. Find the maximum and minimum for $f(x) = x^3 - 12x$

The derivative is $f'(x) = 3x^2 - 12$

Set the derivative equal to 0

$3x^2 - 12 = 0$ giving $x = -2$ or $x = 2.$

Now, we have to determine the sign of $f'(x)$ on either side of -2 or 2.

Using -2.5, and -1.5 gives $f'(-2.5) = 3(-2.5)^2 - 12 = 6.75 > 0$

and $f'(--1.5) = 3(-1.5)^2 - 12 = -5.25 < 0$

The slope is positive on the left side of -2 and negative on the right side, indicating a maximum.

Thus, the function has a maximum of $f(-2) = (-2)^3 - 12(-2) = 16$.

Next using 1.5 and 2.5 we have

$$f'(1.5) = 3(1.5)^2 - 12 = -5.25 < 0$$

and $f'(2.5) = 3(2.5)^2 - 12 = 6.75 > 0$

The slope is negative on the left side of 2 and positive on the right side, indicating a minimum.

Thus, the function has a minimum of $f(2) = 2^3 - 12(2) = -16$.

Exercise 5(a)

Find all stationary values of each function, and determine whether each stationary value is a maximum or a minimum.

1. $f(x) = x^2 - 3x + 2$ 2. $f(x) = 5 + 6x - x^2$

3. $f(x) = 3x - x^3$ 4. $f(x) = 4x - 3x^3$

5. $f(x) = x^3 - 3x + 7$ 6. $f(x) = x^2(2 - x)$

7. $f(x) = 2x^3 - x^2 - 8x + 3$ 8. $f(x) = x(3x^2 + 2x)$

9. $f(x) = 2x^3 - 11x^2 + 12x - 5$ 10. $f(x) = x^3(1 - 8x)$

You can use the second derivative to decide whether a stationary value is a minmum or a maximum.

Second Derivative Test

If $f'(c) = 0$, then

1. $f(c)$ is a local maximum of f if $f''(c)$ is negative.

2. $f(c)$ is a local minimum of f if $f'(c)$ is positve.

Note that it is possible for $f''(c)$ to be zero. In such cases, use the first derivative test.

Example

Find all stationary values of each function. Use the second derivative test to decide whether each stationary value is a local maximum or local minimum.

$$f(x) = x^3 - 5x^2 + 3x + 2$$

Begin by finding the derivative of f.

$$f'(x) = 3x^2 - 10x + 3$$

Next, set the derivative to 0.

$$3x^2 - 10x + 3 = 0$$

$$(3x - 1)(x - 3) = 0$$

giving $x = 1/3$ or $x = 3$

The second derivative is $f''(x) = 6x - 10$

Evaluating $f''(x)$ at $1/3$ gives $f''(1/3) = 6(1/3) - 10 = -8 < 0$, indicating a maximum

So, f has a maximum of $f(1/3) = (1/3)^3 - 5(1/3)^2 + 3(1/3) + 2 = 67/27$

Also, when $x = 3$ then $f''(3) = 6(3) - 10 = 8 > 0$, indicating a minimum.

So, f has a minimum $f(3) = 3^3 - 5(3^2) + 3(3) + 2 = -7$.

Exercise 5(b)

Find all stationary values of each function. Use the second derivative test to decide whether each stationary value is a maximum or a minimum.

1. $f(x) = 3x^2 - 4x + 8$ 2. $f(x) = 8 - 6x - x^2$

3. $f(x) = x^2 - 9$ 4. $f(x) = 4 - x^2$

5. $f(x) = x^2 - 2x$ 6. $f(x) = x^3 - 3x$

7. $f(x) = 27x - x^3$ 8. $f(x) = x^3 - 3x^2 + 5$

9. $f(x) = x^3 - 6x^2 + 8$ 10. $f(x) = 4x^3 - 3x$

6 Applications of Maxim and Minima

The solution of many problems in life may require finding a maximum or a minimum. Your knowledge of stationary values and the method of determining a maximum and a minimum may be useful tools for solving such problems.

Examples

1. Thin metal is used to make cylindrical cans which are to hold $1,200 \text{ cm}^3$ of fruit juice. What should be the radius of the can if they are to use the least amount of metal?

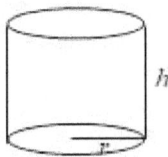

We need to find the relationship between A and r that will give the least area of thin metal. Let r and h be the radius and height of the can respectively.

The area consists of a top and a bottom, each of which is a circle and a curved surface area. The area of the top and bottom of the can is $2\pi r^2$ and area of the curved surface is $2\pi rh$.

Thus, the area of the metal is given by

$$A = 2\pi r^2 + 2\pi rh$$

We cannot differentiate until we have removed one of the variables. Since the volume of the can is 1,200, we can express h in terms r as follows.

$$\pi r^2 h = 1200 \text{ giving } h = 1200/\pi r^2$$

We now substitute this expression for h into the equation for A to get

$$A = 2\pi r^2 + \frac{2400}{r}$$

There are no restrictions on r other than that it must be a positive number greater than 0.

Now, we have to find dA/dr

$$\frac{dA}{dr} = 4\pi r - \frac{2400}{r^2}$$

Setting dA/dr to 0 and solving this equation gives

$$4\pi r - \frac{2400}{r^2} = 0$$

$$r^3 = \frac{600}{\pi}$$

$$r \approx 5.8$$

The only critical number is 5.8.

The second derivative gives

$$\frac{d^2A}{dr^2} = 4\pi + \frac{2400}{r^3}$$

Notice that d^2A/dr^2 is positive for all positive values of r, giving a minimum area when r = 5.8 Thus, the least volume of the can has radius 5.8 cm.

2. An open box is to be made by cutting a square from each corner of 8 cm by 5 cm sheet of metal and then folding up the sides. Find the maximum volume of the box.

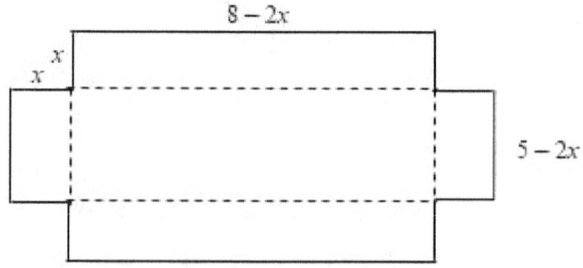

Let x represents the length of a side of the square that is cut from each corner. Then the length of the box is 8 − 2x, and the width is 5 − 2x as shown in the diagram. The depth of the box is x.

The volume of the box is given by Volume = length × width × depth.

So, $V = x(8 - 2x)(5 - 2x)$

$$= 40x - 26x^2 + 4x^3$$

Now, $V'(x) = 40 - 52x + 12x^2$

Set the derivative to 0, and solve the resulting equation.

$12x^2 - 52x + 40 = 0$

or $(3x - 10)(x - 1) = 0$

So, $3x - 10 = 0$ or $x - 1 = 0$ giving $x = 10/3$ or $x = 1$.

The second derivative is $V''(x) = -52 + 24x$. Using the second derivative test, we have

$V''(10/3) = -52 + 80 = 28 > 0$ and $V''(1) = -52 + 24 = -28 < 0$.

The results indicate that the box will have maximum volume when $x = 1$. So the maximum volume will be $V(1) = 40 - 26 + 4 = 18$ cm^3.

Using maxima and minima to solve problems involve the following steps.

1. If possible, sketch a diagram and label its various parts.

2. Determine what the variables are and how they are related.

3. Decide what quantity needs to be maximized or minimized.

4. Write an expression for the quantity to be maximized or minimized in only one variable.

5. Find the maximum or minimum values of the variable you are using.

6. Solve the problem.

Exercise 6(a)

1. A cuboid box is to have a volume of 576 cm^3. The length is twice the breadth. Find the height, if the surface area is to be as small as possible.

2. A metal basket is made in the form of a cylinder with an open top. Its height is h cm and its radius r cm. It is to contain 3,000 cm^3. What is the radius, if it is to consist of the smallest possible amount of metal?

3. A cardboard box is to be made in the form of a cuboid with square cross-section. Its volume must be 8000 cm^3. Let x be the side of the square, and z cm be the length. What value of x will use the least cardboard?

4. A closed cylindrical can is made of a tin plate. If the volume of the can is 64 cm^3, find the radius of the can with the least possible surface area.

5. A manufacturer of tin cans wishes to produce a closed cylindrical can of volume 2 liters. Find the dimensions of the can with the least possible surface area.

6. A metal can is made in the form of a cylinder with an open top. Its height is h cm and its radius r cm. It is to contain 10 cm^3 of liquid. What is the radius, if it is to consist of the smallest possible amount of metal?

7. A right- angled triangle has a hypotenuse of 9 m. Find the maximum area, as the other two sides vary.

8. Find the maximum possible area of a rectangle whose perimeter is 32 cm. What are the dimensions?

9. A three- sided fence is to be built by a farmer next to a straight section of a river, which forms the fourth side of a rectangular field. If there is 200 m of fencing available, find the maximum enclosed area and the dimensions of the corresponding enclosure.

10. A rectangular fence is to be built by a farmer. If the enclosed area is to equal 900 m^2, find the minimum perimeter and the dimensions of the corresponding enclosure.

11. An open box is made by cutting a square region from each corner of a sheet of metal 8 cm by 5 cm and folding up the sides. Find the length of the side of each square region that must be cut so that the volume of the box will be maximum.

12. An open box is made by cutting a square region from each corner of a sheet of metal 12 cm square and folding up the sides. Find the length of the side of each square region that must be cut out so that the volume of the box will be maximum.

13. The bottom of a tank of height h cm is a square of side x m and the tank is open at the top. It is designed to hold 4 m^3 of liquid. Express in terms of x

the total area of the bottom and the four sides of the tank. Find the value of x for which the area is minimum.

14. An open tank is to be constructed with a square horizontal base and vertical sides. The capacity of the tank is to be 400 m^3. The cost of the material for the sides is \$5 per square meter and the base is \$3 per square meter. Find the minimum cost of the material, and give the corresponding dimensions of the tank.

15. A fence must be built in a large field to enclose a rectangular area of 10,000 m^2. One side of the area is bounded by an existing fence; no fence is needed there. Material for the fence cost \$ 2 per meter for the ends and \$ 1.50 per meter for the side opposite the existing fence. Find the cost of the least expensive fence.

16. A closed box with a square base is to have a volume of 6750 cm^3. The material for the top and bottom of the box cost \$ 3 per square centimeter, while the material for the sides cost \$ 1.50 per square centimeters. Find the dimensions of the box that will lead to the minimum total cost. What is the minimum total cost?

17. A company manufactures cylindrical metal containers with volume of 16 m^3. The top and bottom of each container is made of a material that cost \$ 2 per square meter, while the material for the side cost \$1per square meter. Find the radius, height and cost of the least expensive container.

Curve Sketching

The concepts you learned in the preceding section would help you sketch graphs of functions. Curve sketching may be done with the following steps.

1. Find the intercept on the x- and y- axes.

2. Find all possible local maxima and local minima.

3. Find the behavior of the function (if necessary) as x becomes very large or very small.

4. Plot the intercepts, the maximum and minimum points and other points as needed.

5. Finally, connect the points with a smooth curve.

Example

(a) Sketch the graph of $f(x) = x^3 - 3x^2$.

Begin by finding the x- and y- intercepts.

To find the y- intercept we replace x with 0.

Now $f(0) = 0$, so the y- intercept is $(0, 0)$.

To find the x- intercept we need to solve the equation $f(x) = 0$. That is

$$x^3 - 3x^2 = 0$$

Solving this equation we get $x = 0$ or $x = 3$.

The x- intercepts are $(0, 0)$ and $(3, 0)$.

Next find all possible maxima or minima.

Find the derivative of the function and equate it to zero.

$$f'(x) = 3x^2 - 6x$$

$$3x^2 - 6x = 0$$

Solving this equation gives $x = 0$ or $x = 2$.

So, the graph has turning points at $x = 0$ and $x = 2$.

We would use the second derivative test to determine whether we have a maximum or a minimum at the turning points.

Evaluating $f''(x)$ at $x = 0$ gives $f''(0) = 6(0) - 6 = -6 < 0$, indicating a maximum. So f has a maximum of $f(0) = 0$, and a maximum point at $(0, 0)$.

Again evaluating $f''(x)$ at $x = 2$ gives $f''(2) = 6(2) - 6 = 6 > 0$, indicating a minimum.

So, f has a minimum of $f(2) = 2^3 - 3 \cdot 2^2 = -4$, and a minimum point at $(2, -4)$.

Finally, examine the behavior of the function as x decreases or increases.

As $x \to \pm \infty$, we have $f(x) \approx x^3$.

As $x \to +\infty, f(x) \to +\infty,$

and as $x \to -\infty, f(x) \to -\infty.$

Using these results gives a graph as shown in the figure below.

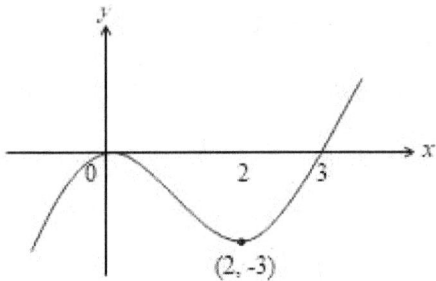

(2, -3)

(b) Sketch the graph of $f(x) = x^3 - x^2 - x + 1.$

Find the x- and y- intercepts.

To find the x- intercept we solve $f(x) = 0.$

$x^3 - x^2 - x + 1 = 0$

Factoring gives $(x - 1)(x^2 - 1) = 0$ or $(x - 1)^2(x + 1) = 0.$ So $x = -1$ or $x = 1.$

The x- intercepts are $(-1, 0)$ and $(1, 0).$

To find the y- intercept we replace x with 0.

Hence, $f(0) = 1.$ So, the y- intercept is $(0, 1).$

Next, find all possible maxima or minima..

$f'(x) = 3x^2 - 2x - 1$

Setting $f'(x)$ to 0 gives $3x^2 - 2x - 1 = 0$

$(3x + 1)(x - 1) = 0$

So, $x = -1/3$ or $x = 1.$

The graph has turning points at $x = -1/3$ and $x = 1.$

Now, find the second derivative. $f''(x) = 6x - 2$

Evaluating $f''(x)$ at $x = -1/3$ gives $f''(-1/3) = 6(-1/3) - 2 = -4 < 0$

So, f has a maximum of $f(-1/3) = (-1/3)^3 - (-1/3)^2 - (-1/3) + 1 = 32/27$

and maximum point at $(-1/3, 32/27)$.

Also, $f''(1) = 6 - 2 = 4 > 0$

So, f has a minimum of $f(1) = 0$, and minimum point at $(1, 0)$.

As $x \to \pm\infty$ we have $f(x) \approx x^3$.

Also, as $x \to -\infty$ $f(x) \to -\infty$

and as $x \to \infty$ $f(x) \to \infty$.

A graph of the function is as shown in the following figure.

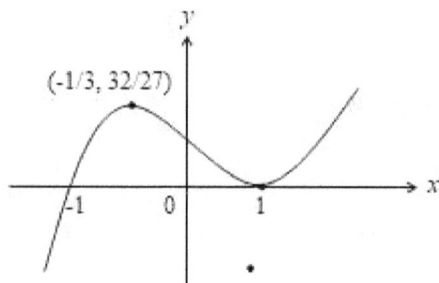

Exercise 6(b)

Sketch the graph of each of the following functions.

1. $y = x^2 - x - 6$ 2. $y = 3 + 2x - x^2$

3. $y = x^3 - 12x$ 4. $y = x^3 - 3x$

5. $y = 3x^2 - x^3$ 6. $y = x^3 - 6x^2$

7. $y = x^3 - 2x^2 + x$ 8. $y = x^2(x - 1)$

9. $y = x^3 - 6x^2 + 9x$ 10. $y = x^4 - 2x^3$

11. $y = 4x^5 - 5x^4$ 12. $y = 4x^3 + 3x^4$

7 Integration

Indefinite Integrals

To integrate a given function $f(x)$ is to find a function $F(x)$ whose derivative is $f(x)$. Such a function is called an antiderivative of $f(x)$. For example, if $f'(x) = 3x^2$, then $F(x) = x^3$ is an antiderivative of f. Since the derivative of a constant is 0, $F(x) = x^3$ is not the only function whose derivative is $f(x) = 3x^2$. For example, $F(x) = x^3 + 2$ and $G(x) = x^3 - 5$ are some antiderivatives of $f(x) = 3x^2$. The antiderivatives of functions differ only by a constant. The antiderivatives of a function f is indicated by

$$\int f(x)\, dx = F(x) + c$$

where c is a constant called an integration constant.

The symbol $\int f(x)\, dx$ is called an indefinite integral, $f(x)$ is called the integrand, and $\int$ is the integral sign. The term dx indicates that $f(x)$ is integrated with respect to x.

Rule for integrating powers of x

For any real number $n \neq -1$,

$$\int x^n\, dx = \frac{x^{n+1}}{n+1} + c$$

That is, to integrate x increase the exponent by 1 and divide by the new exponent, $n+1$.

If $n = -1$ the expression in the denominator is 0, and the above rule cannot be used.

Example

Find each of the following indefinite integrals.

(a) $\int x^7\, dx$

$$\int x^7\, dx = \frac{x^{7+1}}{7+1} + c = \frac{1}{8}x^8 + c$$

(b) $\int x^{-4}\, dx$

$$\int x^{-4} \, dx = \frac{x^{-4+1}}{-4+1} + c = -\frac{1}{3}x^{-3} + c$$

(c) $\int x^{1/2} \, dx$

$$\int x^{1/2} \, dx = \frac{x^{\frac{1}{2}+1}}{\frac{1}{2}+1} + c = \frac{2}{3}x^{3/2} + c$$

Exercise 7(a)

Find the following indefinite integrals.

1. $\int x^2 \, dx$

2. $\int x^5 \, dx$

3. $\int x^8 \, dx$

4. $\int 5 \, dx$

5. $\int x \, dx$

6. $\int x^{-4} \, dx$

7. $\int x^{-3} \, dx$

8. $\int x^{-6} \, dx$

9. $\int x^{2/3} \, dx$

10. $\int x^{3/4} \, dx$

11. $\int x^{-1/2} \, dx$

12. $\int x^{-2/5} \, dx$

The following properties of indefinite integrals hold.

Constant Multiple Rule

If all indicated integrals exist, then for any real number k,

$$\int k \cdot f(x) \, dx = k \cdot \int f(x) \, dx$$

The antiderivative of constant times a function is the constant times the antiderivative of the function.

Example

Find the following indefinite integrals.

(a) $\int 2x^3 \, dx$

$$\int 2x^3 \, dx = 2 \int x^3 \, dx$$

$$= 2 \cdot \frac{1}{4}x^4 + c$$

$$= \frac{1}{2}x^4 + c$$

(b) $\int 3x^{1/2} \, dx$

$$\int 3x^{1/2} \, dx = 3 \int x^{1/2} \, dx$$

$$= 3 \cdot \frac{2}{3}x^{3/2} + c$$

$$= 2x^{3/2} + c$$

Exercise 7(b)

Find each integral.

1. $\int 6x \, dx$ 2. $\int 4x^5 \, dx$

3. $\int 8x^3 \, dx$ 4. $\int \frac{3}{4} x^2 \, dx$

5. $\int \frac{5}{4}x^4 dx$ 6. $\int 3x^{-4} \, dx$

7. $\int \frac{2}{3} x^{-3} \, dx$ 8. $\int 2x^{-1/2} \, dx$

9. $\int 4x^{1/3} dx$ 10. $\int 3x^{-1/4} \, dx$

11. $\int 5x^{3/2} \, dx$ 12. $\int 6x^{-5} \, dx$

Sum or Difference Rule

If all indicated integral exist,

$$\int [f(x) \pm g(x)] \, dx = \int f(x) \, dx \pm \int g(x) \, dx$$

The antiderivative of a sum or difference of functions is the sum or difference of the antiderivative.

Example

Find the following indefinite integrals.

(a) $\int (x^3 + x^2 + 1) \, dx$

$$\int (x^3 + x^2 - 1)\, dx = \int x^3\, dx + \int x^2\, dx - \int dx$$

$$= \frac{1}{4}x^4 + \frac{1}{3}x^3 - x + c$$

(b) $\int (3x^2 - 2x + 4)\, dx$

$$\int (3x^2 - 2x + 4)\, dx = 3 \int x^2\, dx - 2 \int x\, dx + 4 \int dx$$

$$= x^3 - x^2 + 4x + c$$

Exercise 7(c)

Find each integral.

1. $\int (2x^2 + x^3)\, dx$

2. $\int (2x^3 - 5x^4)\, dx$

3. $\int (3x^2 - 2x + 1)\, dx$

4. $\int (3 - 2x + 4x^2)\, dx$

5. $\int (1 - 2x + 6x^2)\, dx$

6. $\int (5x^4 - 3x^2 + 2x)\, dx$

7. $\int x(x^2 - 2)\, dx$

8. $\int x(4x^2 + 3x)\, dx$

9. $\int x^2(2 + 3x^2)\, dx$

10. $\int x(x^4 - 3x)\, dx$

Definite Integrals

If $F(x)$ is any antiderivative of the function f, then the definite integral of f from a to b is given by

$$\int_a^b f(x)\, dx = F(b) - F(a)$$

The number b above the integral sign is called the upper limit of integration, and a is called the lower limit of integration. The symbol $[F(x)]_a^b$ is used to represent $F(b) - F(a)$.

The following properties of definite integrals hold.

1. $\int_a^b kf(x)\, dx = k \cdot \int_a^b f(x)\, dx$, for any real number k.

2. $\int_a^b [f(x) \pm g(x)]\, dx = \int_a^b f(x)\, dx \pm \int_a^b g(x)\, dx$

Example

Find each definite integral.

(a) $\int_1^2 x^3 \, dx$

$$\int_1^2 x^3 \, dx = \left[\frac{1}{4}x^4 + c\right]_1^2$$

$$= \left(\frac{1}{4} \times 2^4 + c\right) - \left(\frac{1}{4} \times 1^4 + c\right)$$

$$= (4 + c) - \left(\frac{1}{4} + c\right)$$

$$= 4 - \frac{1}{4}$$

$$= \frac{15}{4}$$

This example shows that the constant c is not required in this case, because it would be eliminated in the final answer.

(b) $\int_0^2 (3x^2 - 4x^3) \, dx$

$$\int_0^2 (3x^2 - 4x^3) \, dx = [x^3 - x^4]_0^2$$

$$= (2^3 - 2^4) - 0$$

$$= -8$$

Notice that the definite integral represents a number.

Exercise 7(d)

Find the following definite integrals:

1. $\int_{-1}^2 x \, dx$

2. $\int_1^3 x^2 \, dx$

3. $\int_2^4 x^3 \, dx$

4. $\int_{-2}^{-1} x^4 \, dx$

5. $\int_{-1}^1 \frac{3}{2}x^2 \, dx$

6. $\int_0^4 x^{-3} \, dx$

7. $\int_{-2}^{2}(x^2 + x)\,dx$ 8. $\int_{0}^{2}(4 - x^2)\,dx$

9. $\int_{2}^{4}(3x^2 - 2x)\,dx$ 10. $\int_{1}^{2}(x^3 - x)\,dx$

11. $\int_{0}^{2} x^2(1 + 4x)\,dx$ 12. $\int_{-1}^{2}(x^3 + 2)^2\,dx$

Integration by Substitution

Occasionally, we evaluate $\int f(x)\,dx$ by making the substitution $x = g(u)$, and then using the following rule.

$\int f(x)\,dx = \int f[g(u)]g'(u)du$

Example

Find the following.

(a) $\int (3x - 2)^3\,dx$

Let $u = 3x - 2$, so $du = 3dx$. Now substitute u for $3x - 2$ and $(1/3)\,du$ for dx.

$$\int (3x - 2)^3\,dx = \int u^3 \cdot \frac{du}{3}$$

$$= \frac{1}{3}\int u^3\,du$$

$$= \frac{1}{3}\left(\frac{1}{4}u^4\right) + c$$

$$= \frac{1}{12}u^4 + c$$

Substituting $3x - 2$ for u we get

$$\int (3x - 2)^3\,dx = \frac{1}{12}(3x - 2)^4 + c$$

(b) $\int_{-1}^{1} x^2(1 - x^3)\,dx$

With a definite integral, the limits should be changed. The new limits are found as follows.

Let $u = 1 - x^3$, so $du = -3x^2\,dx$.

If $x = 1$, then $u = 1 - 1 = 0$

If $x = -1$, then $u = 1 + 1 = 2$. So,

$$\int_{-1}^{1} x^2 (1 - x^3) \, dx = \int_{2}^{0} x^2 \cdot u \cdot \frac{du}{-3x^2}$$

$$= -\frac{1}{3} \int_{2}^{0} u \, du$$

$$= \left[-\frac{1}{6} u^2 \right]_{2}^{0}$$

$$= 0 + \frac{2}{3}$$

$$= \frac{2}{3}$$

An alternative method is to ignore the limits and evaluate the antiderivative. Then substitute the original limits.

Exercise 7(e)

Use substitution to find each indefinite integral.

1. $\int (3x + 2)^3 \, dx$

2. $\int (2x - 1)^{-2} \, dx$

3. $\int \sqrt[3]{x - 2} \, dx$

4. $\int x(x^2 + 3)^2 \, dx$

5. $\int x(x^2 + 1)^4 \, dx$

6. $\int x^2 (2x^3 + 3)^4 \, dx$

7. $\int x\sqrt{1 - x^2} \, dx$

8. $\int x^2 \sqrt{x^3 - 2} \, dx$

9. $\int 4x^2 \sqrt{1 - x^3} \, dx$

10. $\int x^2 \sqrt{x^3 + 4}\, dx$

Evaluate each definite integral.

11. $\int_0^1 (1 - x)^4\, dx$

12. $\int_2^7 \sqrt{x + 2}\, dx$

13. $\int_0^2 x^2 \sqrt{x^3 + 1}\, dx$

14. $\int_{-1}^2 \frac{1}{(2x+1)^2}\, dx$

15. $\int_1^2 \frac{1}{(3x-2)^2}\, dx$

16. $\int_0^1 \frac{x^2}{(x^4+1)^3}\, dx$

8 Applications of Definite Integrals

Area under a Curve

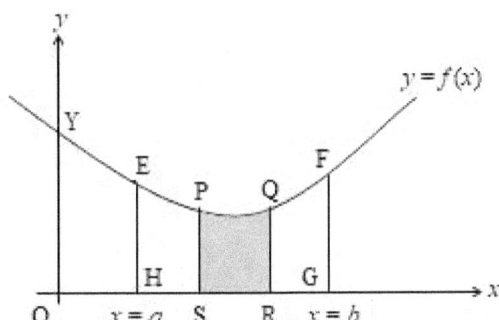

Consider the area EFGH bounded by the curve $y = f(x)$, the x-axis and the two vertical lines EH and FG through $x = a$ and $x = b$ respectively.

To evaluate the area EFGH, we consider the total area A, between the curve and the x-axis from OY up to some arbitrary point P on the curve with coordinates (x, y). If we move the point P to the point Q with coordinates $(x + \Delta x, y + \Delta y)$ the right-hand boundary of A moves from PS to QR, resulting in the increase ΔA in the area A. If the area under the arc PQ is approximated by a rectangle then ΔA is approximately $y\Delta x$. That is

$$\Delta A \approx y\,\Delta x.$$

Dividing each side by Δx gives

$$\frac{\Delta A}{\Delta x} \approx y$$

The limit of $\Delta A / \Delta x$ as $\Delta x \to 0$ is

$$\frac{dA}{dx} = y$$

so $A = \int y\,dx$. The symbol $\int y\,dx$ represent the area between the curve and the x-axis up to the point P. The required area is the difference between the area OYFG and OYEH, which is equal to the definite integral of $f(x)$ from a to b, denoted by

$$A = \int_a^b f(x)\, dx$$

Example

(a) Find the area of the region between the x- axis and the graph of $y = 3x$ from $x = 1$ and $x = 2$.

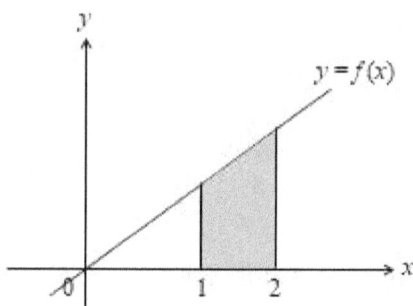

The region is shown in the figure above. The area of the shaded region is given by

$$Area = \int_1^2 y\, dx$$

$$= \int_1^2 3x\, dx$$

$$= \left[\frac{3}{2}x^2\right]_1^2$$

$$= 6 - \frac{3}{2}$$

$$= \frac{9}{2}$$

The required area is 9/2.

You can use the formula of the area of a triangle $A = \frac{1}{2}\, bh$ to confirm this result. The area of the larger triangle minus the area of the smaller triangle gives the area A of the curve from $x = 1$ to $x = 2$.

$$A = \frac{1}{2} \cdot 2 \cdot 6 - \frac{1}{2} \cdot 1 \cdot 3 = 9/2$$

(b) Find the area between the x- axis and the graph of $y = x^2 - 5x + 6$ from 2 to 3.

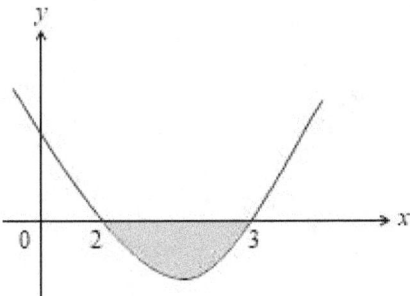

The region shown in the figure above lies below the x- axis. The area is given by

$$\int_2^3 (x^2 - 5x + 6)\, dx = \left[\tfrac{1}{3}x^3 - \tfrac{5}{2}x^2 + 6x\right]_2^3$$

$$= \left(9 - \tfrac{45}{2} + 18\right) - \left(\tfrac{8}{3} - 10 + 12\right)$$

$$= \tfrac{9}{2} - \tfrac{14}{3}$$

$$= -\tfrac{1}{6}$$

Sometimes the integral gives a negative, especially when the required region is below the x- axis.

The required area is $|1/6|$.

(c)   Find the area between the x- axis and the graph of $f(x) = x^3 - 6x^2 + 8x$ from $x = 0$ to $x = 4$.

Start by sketching the graph of $f(x) = x^3 - 6x + 8x$.

A graph of the function is below.

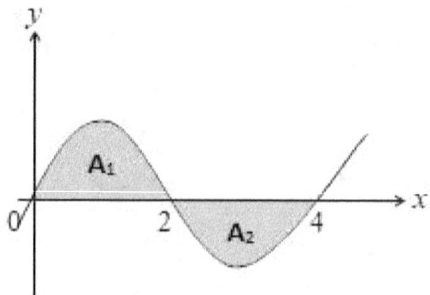

Part of the region is below the x- axis. The definite integral over the region below the x- axis will have a negative value. To find the area, integrate the negative and positive portions separately.

$A_1 = \int_0^2 (x^3 - 6x^2 + 8x)\, dx$

$\qquad = \left[\frac{1}{4}x^4 - 2x^3 + 4x^2\right]_0^2$

$\qquad = (4 - 16 + 16) - 0$

$\qquad = 4$

$A_2 = \int_2^4 (x^3 - 6x^2 + 8x)\, dx$

$\qquad = \left[\frac{1}{4}x^4 - 2x^3 + 4x^2\right]_2^4$

$\qquad = (64 - 128 + 64) - (4 - 16 + 16)$

$\qquad = -4$

To find the area add the absolute value of A_1 and of A_2.

The total area is $4 + |-4| = 8$.

Exercise 8(a)

Find the area of the region between the x- axis and the graph of:

1. $y = 5x$ from $x = 1$ to $x = 4$

2. $y = x^2$ from $x = 1$ to $x = 3$

3. $y = 2x^2 - x + 1$ from $x = -1$ to $x = 2$

4. $y = x^2 + 3$ from $x = -1$ to $x = 2$

5. $y = (x - 1)(x - 3)$ from $x = 0$ to $x = 3$

6. $y = \dfrac{1}{x^2}$ from $x = 1$ to $x = 8$

Find the area between the x- axis and the graph of:

7. $y = x(3 - x)$ from $x = 0$ to $x = 4$

8. $y = -x^3$ from $x = -2$ to $x = 2$

9. $y = x^2(x - 1)$ from $x = 0$ to $x = 2$

10. $y = 4 - x^2$ from $x = 0$ to $x = 3$

11. $y = x^3 - 4x$ from $x = -2$ to $x = 2$

12. Find the area above the x- axis and below $y = 6 - x - x^2$

The Area between Two Curves

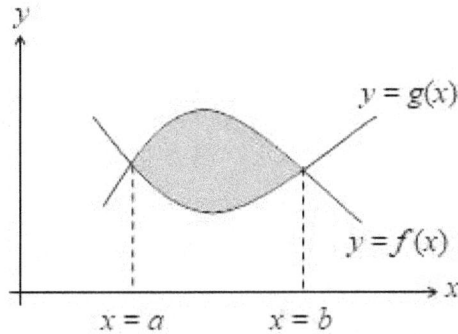

The figure above shows the graphs of $y = f(x)$ and $y = g(x)$. The area between the two curves between $x = a$ and $x = b$ (see shaded region) is the area under the graph of $f(x)$ minus the area under the graph of $g(x)$. That is, the area between the graphs is given by

$$\int_a^b f(x)\,dx - \int_a^b g(x)\,dx$$

which is written briefly as

$$\int_a^b [f(x) - g(x)]\, dx$$

The limits of integration correspond to the x- coordinates of the points of intersection. To find the limits, we set the two functions equal and solve for x.

Example

(a) Find the area between the curves $y = x^2$ and $y = 2x$.

The graphs of the functions are shown in the figure below.

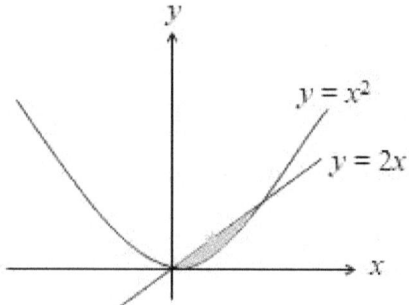

Begin by finding the limits of the integration. Set x^2 equal to $2x$.

$$x^2 = 2x$$

$$x(x - 2) = 0$$

giving $x = 0$ or $x = 2$.

The area between the two curves is given by

$$\int_0^2 (2x - x^2)\, dx = \left[x^2 - \frac{1}{3}x^3 \right]_0^2$$

$$= \left(4 - \frac{8}{3} \right) - 0$$

$$= \frac{4}{3}$$

The area is 4/3.

(b) Find the area of the region enclosed by $y = x(x - 2)$ and $y = x(4 - x)$.

Solve the equation

$$x(x - 2) = x(4 - x)$$

$$2x^2 - 6x = 0 \text{ or } 2x(x - 3) = 0 \text{ giving } x = 0 \text{ or } x = 3$$

The graphs of the functions are shown in the figure below.

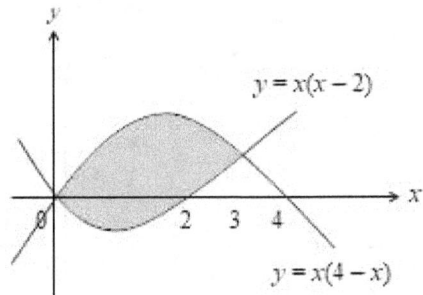

The area between the two curves is given by

$$\int_0^3 [x(4 - x) - x(x - 2)] = \int_0^3 (6x - 2x^2) \, dx$$

$$= \left[3x^2 - \frac{2}{3}x^3 \right]_0^3$$

$$= (27 - 18) - 0$$

$$= 9$$

The area is 9.

Exercise 8(b)

Find the area of the segment cut off from each of the following curves by the given straight lines.

1. $y = x^2 - 3x - 4$, $y = 6$ 2. $y = x^2 + 2x - 3$, $y = 5$

3. $y = -x^2 + x + 6$, $y = -6$ 4. $y = x(x - 1)$, $y = x$

5. $y^2 = x$, $y = x$ 6. $y = 8 - 2x - x^2$, $y + x - 2 = 0$

7. $y = x^2 - x - 6,\ y = 2x - 2$ 　　　　8. $y = x^2 + 1,\ y = 5$

Find the area between the curves.

9. $y = 4 - x^2,\ y = x(x - 2)$ 　　　　10. $y = x(x - 2),\ y = x(4 - x)$

11. $y = 2x^2 - 4x,\ y = x^2 - 3x$ 　　　　12. $y^2 = x,\ y = x^3$

Solids of Revolution

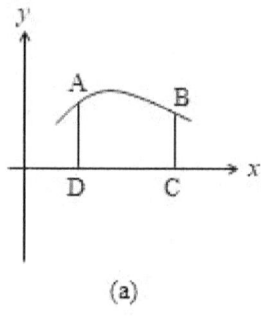

(a)

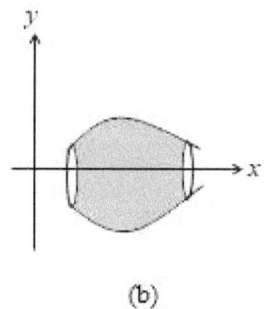

(b)

Many three dimensional solids can be generated by revolving a curve about the x-axis or the y-axis. The solid generated is called solid of revolution. Figure (b) shows the solid generated when AB (see Figure (a)) is rotated through 360° about the DC on the x-axis.

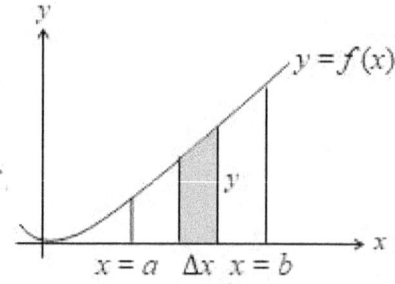

Consider a small element of length y and thickness Δx, between $x = a$ and $x = b$. When this element is revolved about the x-axis, it generates a cylindrical disc of volume $\pi y^2\,\Delta x$. Dividing the region between $x = a$ and $x = b$ into n smaller discs and adding up their volumes, the volume, V, of the solid would be approximated by

$$V = \sum_{i=1}^{n} \pi y^2 \, \Delta x$$

The solid of revolution is the limit of this sum as $\Delta x \to 0$. The limit is a definite integral given by

$$\int_a^b \pi y^2 \, dx$$

Example

(a) Find the volume of the solid generated by rotating about the x-axis the area under $y = x^2$, $x \geq 0$, from $x = 1$ to $x = 2$.

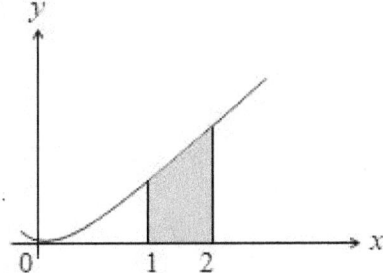

The volume V is given by

$$V = \int_1^2 \pi y^2 \, dx$$

$$= \int_1^2 \pi x^4 \, dx$$

$$= \left[\frac{1}{5} x^5 \right]_1^2$$

$$= \left(\frac{1}{5} \cdot \pi \cdot 2^5 \right) - \left(\frac{1}{5} \cdot \pi \cdot 1^5 \right)$$

$$= \frac{31}{5} \pi$$

The volume of the solid is $31\pi/5$ cubic unit.

(b) Find the volume of the solid generated by rotating about the y-axis the area in the first quadrant enclosed by $y = x^3$, $y = 1$, $y = 8$ and the y-axis.

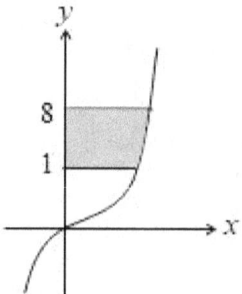

The volume is given by

$$V = \int_1^8 \pi x^2 \, dy$$

$$= \int_1^8 \pi y^{2/3} \, dx$$

$$= \left[\frac{3}{5} \pi y^{5/3} \right]_1^8$$

$$= \left(\frac{3}{5} \cdot \pi \cdot 32 \right) - \left(\frac{3}{5} \cdot \pi \cdot 1 \right)$$

$$= \frac{93}{5}$$

The volume of the solid is $93\pi/5$ cubic unit.

Exercise 8(c)

1. Find the volumes enclosed when the following curves between the limits shown are rotated about the x-axis.

(a) $y = x$, $x = 1$, $x = 2$

(b) $y = 3x$, $x = 0$, $x = 4$

(c) $y = x^2 + 1$, $x = 0$, $x = 1$

(d) $y = 2x - 3$, $x = 3$, $x = 4$

(e) $y = x(x - 2)$, $y = 0$

2. Find the volumes enclosed when the following curves between the limits shown are rotated about the y-axis.

(a) $y = x^2 + 1, x = 0, y = 4$

(b) $y = 1 - x^3, x = 0, y = 0$

(c) $y = 2x - 4, y = 2, x = 0$

(d) $y = x^2, y = 1, y = 4, x = 0$

(e) $y^2 = x - 2, x = 0, y = 0, y = 3$

3.   By considering the revolution of the line $y = rx/h$ about the x-axis between the limits $x = 0$ and $x = h$ (where r and h are constants), find the volume of a cone radius r and height h.

4. The equation of a circle radius r, is $x^2 + y^2 = r^2$. By finding the volume of revolution between $x = -r$ and $x = +r$, find the volume of a sphere.

Answers to Exercises

Exercise 1

1. 9   2. 6   3. 4   4. -5

5. 3   6. -2 7. 12   8. 4

9. 5   10. 3   11. 3/8   12. 6

13. -3 14. 8   15. -2 16. 1

17. 2   18. $6a+1$ 19. 2   20. $4a$

Exercise 2(a)

1. 1   2. 5   3. $6x$ 4. $-2x$ 5. 0   6. $-x$

7. $2x^2$ 8. $-6x$ 9. $6x^2$ 10. $6x$ 11. $-6x$ 12. $2x$

Exercise 2(b)

1. -1 2. 0   3. $4x^3$ 4. $6x^5$ 5. $3x^{-4}$ 6. $-x^{-2}$

7. $-8x^{-9}$ 8. $-\frac{3}{4}x^{-7/4}$ 9. $\frac{1}{2}x^{3/2}$ 10. $-\frac{3}{2}x^{-5/2}$

11. $\frac{2}{3}x^{-5/3}$ 12. $-\frac{1}{3}x^{-2/3}$

Exercise 2(c)

1. -1 2. $12x^3$.   3. $-10x^4$ 4. $\frac{3}{4}x^2$ 5. $4x^5$ 6. $2x^{-6}$

7. $-6x^{-4}$ 8. $12x^{-4}$ 9. $-6x^{-3}$ 10. $x^{-2/3}$ 11. $-3x^{1/2}$

12. $3x^{-1/2}$ 13. $4x^{-5/3}$ 14. $-6x^{-7/4}$ 15. $-6x^{-5/3}$ 16. $-4x^{-7}$

Exercise 2(d)

1. $6x^2 + 12x^3$ 2. $15x^4 - 2$ 3. $-4t^3 + 5t^4$ 4. $2t^{-2}$

5. $12t^2 + 3$ 6. $20x^3 - 15x^4$ 7. $5x^{-6}$ 8. $6x^2 + 6x^{-3}$

9. $12t^3 - 4$ 10. $4t - 3$ 11. $12x + 5$ 12. $t^2 + t^{-3}$

13. $x^{-2/3} + x^{-3/2}$ 14. $-3t^{-7/4} + t^2$ 15. $4x^{-5/3}$ 16. $-3x^{-4}$

17. $3 + \frac{1}{2}x^{-3/2} - x^{-2}$ 18. $-10x^{-3} + \frac{3}{2}x^{-5/2}$ 19. $2t - 12t^{-4}$

20. $6 - 4x$ 21. $6x - 12$ 22. $1 - \frac{4}{3}x$ 23. $3 + x^{-2}$ 24.

$-5t^{-2} + 6t^{-3}$

Exercise 2(e)

1. $20x^4 - 9x^2$ 2. $4x + 5$ 3. $18x^2 + 6x$ 4. $12x - 1$

5. $8x^3 - 9x^2$ 6. $6 - 2x - 6x^2$ 7. $3x^{1/2} + \frac{3}{2}x^{-1/2}$

8. $2x + 8x^3$ 9. $\frac{2}{3}x^{-2/3} + \frac{4}{3}x^{1/3}$ 10. $-6x^{-2} + 18x^{-3}$

11. $7x^6 + 4x^3 - 3x^2$ 12. $18x^2 - 30x + 4$

Exercise 2(f)

1. $\dfrac{1}{(x+1)^2}$ 2. $\dfrac{2}{(x-1)^2}$ 3. $\dfrac{-2x}{(x^2-1)^2}$ 4. $\dfrac{-12x}{(3x^2-1)^2}$

5. $\dfrac{-3x^2-6}{(x^2-2)^2}$ 6. $\dfrac{6x^2-42x+10}{(2x-7)^2}$ 7. $\dfrac{-1}{\sqrt{x}(1+\sqrt{x})^2}$

8. $\dfrac{4x}{(1-x^2)^2}$ 9. $\dfrac{x^2+2x}{(x+1)^2}$ 10. $\dfrac{-2x}{(x^2-1)^2}$

11. $\dfrac{2x^2+6x}{(2x+3)^2}$ 12. $\dfrac{-1}{2\sqrt{x}(\sqrt{x}-1)}$

Exercise 2(g)

1. $3(x-5)^2$ 2. $15(3x-4)^4$ 3. $24x(1-2x^2)^5$ 4. $8x(x^2+1)^3$

5. $21x^2(x^3-2)$ 6. $-6(2x+1)^{-4}$ 7. $8(3-2x)^{-5}$

8. $-8(4x+3)^{-3}$ 9. $4x(1-2x^2)^{-2}$ 10. $x(x^2-1)^{-1/2}$

11. $-2(3x+2)^{-5/3}$ 12. $3x(2x^2-2)^{-1/4}$

Exercise 2(h)

1. $4(x+3)^3$ 2. $9(3x-2)^2$ 3. $20(2x^2+3)^4$ 4. $x(x^2+1)^{-1/2}$

5. $-2x^2(2x^3-1)^{-4/3}$ 6. $-4x(3+x^2)^{-3}$ 7. $-6x(1-x^2)^2$

8. $-18x(3x^2-1)^{-4}$ 9. $-12x(1-x^2)^5$ 10. $\dfrac{-8x}{(x^2-5)^5}$

11. $\dfrac{-2x}{\sqrt{(2x^2+3)^3}}$ 12. $\dfrac{-2x^2+3}{3\sqrt[3]{(x^3+3x)^4}}$

Exercise 3

1. $12x^2+2$ 2. $2-6x$ 3. $6x^{-3}$ 4. $12x^{-5}$ 5. $12x^{-4}$

6. $-48x^{-5}$ 7. $\dfrac{2}{9\sqrt[3]{x^5}}$ 8. $\dfrac{4}{9\sqrt[3]{x^7}}-\dfrac{3}{2\sqrt{x^5}}$ 9. $6x(5x^3+2)$

10. $\dfrac{-1}{2\sqrt{x^3}}$ 11. $12(x-3)(2x-3)$ 12. $2(3x+4)$ 13. $\dfrac{-2}{(x+1)^3}$

14. $\dfrac{4(1-3x^2)}{(1-x^2)^3}$ 15. $\dfrac{3}{4\sqrt{(1+x)^5}}$ 16. $\dfrac{4+2x-3x^2}{2\sqrt{(x-1)^5}}$

Exercise 4(a)

1. $16\text{ cm}^2\text{ s}^{-1}$ 2. $4.8\pi\text{ cm}^3\text{ s}^{-1}$ $7.2\pi\text{ cm}^3\text{ s}^{-1}$ 3. $3/10\pi\text{ cm s}^{-1}$

4. $2/45\pi\text{ cm s}^{-1}$ 5. $1/8\text{ cm s}^{-1}$ $6\text{ cm}^2\text{ s}^{-1}$ 6. $1000\pi\text{ cm}^3\text{ s}^{-1}$

7. $1/200\pi\text{ m s}^{-1}$ 8. $1/2\pi\text{ cm s}^{-1}$ 9. $1/20\pi\text{ cm s}^{-1}$ 10. $1/5\pi\text{ cm s}^{-1}$

Exercise 4(b)

1. 2 cm   2. 3.84 cm^2 3. 5.07π cm^3 4. 6 %   5. 2.5 %

6. 2.04 m   7. 3 %   8. 4 %   6 %   9. 0.72π cm   10. 4/3 %

11. 2,325   12. 3.011   13. 2.000125   14. 4.15   15. 3.00185

16. 3.074   17. 12.125   18. 2.05   19. 10.0233

Exercise 5(a)

1. Local minimum of $-1/4$ at $x = 3/2$

2. Local maximum of 14 at $x = 3$

3. Local maximum of 2 at $x = 1$;

   Local minimum of -2 at $x = -1$

4. Local maximum of $16/9$ at $x = 2/3$;

   Local minimum of $-16/9$ at $x = -2/3$

5. Local minimum of 5 at $x = 1$;

   Local maximum of 9 at $x = -1$

6. Local minimum of 0 at $x = 0$;

   Local maximum at $32/27$ at $x = 4/3$

7. Local maximum of 8 at $x = -1$;

   Local minimum of $-127/27$ at $x = 4/3$

8. Local minimum of 0 at $x = 0$

   Local maximum of $32/243$ at $x = -4/9$

9. Local maximum of $-35/27$ at $x = 2/3$;

Local minimum of -14 at $x=3$

10. Local minimum of 0 at $x=0$;

Local maximum of $1/256$ at $x=1/4$

Exercise 5(b)

1. Local minimum of $20/3$ at $x=2/3$

2. Local maximum of 17 at $x=-3$

3. Local minimum of -9 at $x=0$

4. Local maximum of 4 at $x=0$

5. Local minimum of -1 at $x=1$

6. Local maximum of 2 at $x=-1$

7. Local maximum of 54 at $x=3$;

Local minimum of -54 at $x=-3$

8. Local maximum of 5 at $x=0$;

Local minimum of 1 at $x=2$

9. Local maximum of 8 at $x=0$;

Local minimum of -24 at $x=4$

10. Local minimum of -1 at $x=1/2$;

Local maximum of 1 at $x=-1/2$

Exercise 6(a)

1. 8 cm   2. 9.85   3. 20   4. 2.17

5. Radius is 6.83 cm; height is 93.24 cm   6. 1.47 cm   7. 81/4

8. 8 cm by 8 cm   9. 5000 m^2 50 cm by 100 cm

10. 1200 cm 30 cm by 60 cm   11. 1 cm   12. 2 cm   13. 2 cm

14. $ 3600   20 cm by 20 cm by 6 cm   15. $ 600

16. 15 cm by 15 cm by 30 cm $4050

17. 1.37 cm   23. 43 cm $ 70.30

Exercise 7(a)

1. $\frac{1}{3}x^3 + c$ 2. $\frac{1}{6}x^6 + c$ 3. $\frac{1}{9}x^9 + c$ 4. $5x + c$ 5. $\frac{1}{2}x^2 + c$

6. $-\frac{1}{3}x^{-3} + c$ 7. $-\frac{1}{2}x^{-2} + c$ 8. $-\frac{1}{5}x^{-5} + c$ 9. $\frac{3}{5}x^{5/3} + c$

10. $\frac{4}{7}x^{7/4} + c$ 11. $2x^{1/2} + c$ 12. $\frac{5}{3}x^{3/5} + c$

Exercise 7(b)

1. $3x^2 + c$ 2. $\frac{2}{3}x^6 + c$ 3. $2x^4 + c$ 4. $\frac{1}{4}x^3 + c$ 5. $\frac{1}{4}x^5 + c$

6. $-x^{-3} + c$ 7. $-\frac{1}{3}x^{-2} + c$ 8. $4x^{1/2} + c$ 9. $3x^{4/3} + c$

10. $4x^{3/4} + c$ 11. $2x^{5/2} + c$ 12. $-\frac{3}{2}x^{-4} + c$

Exercise 7(c)

1. $\frac{2}{3}x^3 + \frac{1}{4}x^4 + c$ 2. $\frac{1}{2}x^4 - x^5 + c$ 3. $x^3 - x^2 + x + c$

4. $3x - x^2 + \frac{4}{3}x^3 + c$ 5. $x - x^2 + 2x^3 + c$ 6. $x^5 - x^3 + x^2 + c$

7. $\frac{1}{4}x^4 - x^2 + c$ 8. $x^4 + x^3 + c$ 9. $\frac{2}{3}x^3 + \frac{3}{5}x^5 + c$

10. $\frac{1}{6}x^6 - x^3 + c$

Exercise 7(d)

1. 3/2   2. 26/3   3. 80   4. 31/5   5. 1   6. $-1/32$

7. 28/3   8. 16/3   9. 44   10. 9/4   11. 56/3   12. 41/4

Exercise 7(e)

1. $\dfrac{1}{12}(3x+2)^4 + c$ 2. $-\dfrac{1}{(2x-1)} + c$

3. $\dfrac{3}{4}\sqrt[3]{(x-2)^2} + c$ 4. $\dfrac{1}{6}(x^2+3)^3 + c$

5. $\dfrac{1}{10}(x^2+1)^5 + c$ 6. $\dfrac{1}{30}(2x^3+3)^5 + c$

7. $-\dfrac{1}{3}\sqrt{(1-x^2)^3} + c$ 8. $\dfrac{2}{9}\sqrt{(x^3-2)^3} + c$

9. $-\dfrac{8}{9}\sqrt{(1-x^3)^3} + c$ 10. $\dfrac{2}{9}\sqrt{(x^3+4)^3} + c$

11. 1/5   12. 38/3   13. 52/3   14. 2/5

15. 1/4   16. 3/32

Exercise 8(a)

1. 60   2. 26/3   3. 15/2   4. 12   5. 8/3   6. 7/8

7. 19/3   8. 8   9. 3/2   10. 23/3   11. 8   12. 56/3

Exercise 8(b)

1. 343/6   2. 36   3. 343/3   4. 4/3   5. 1/6   6. 125/6

7. 125/6   8. 16   9. 9   10. 9   11. 1/6   12. 5/12

Exercise 8(c)

1. (a) π (b) 192π (c) $28\pi/15$ (d) $77\pi/3$ (e) $56\pi/15$

2. (a) $9\pi/2$ (b) $3\pi/5$ (c) 18π (d) $15\pi/2$ (e) $483\pi/5$

3. $\frac{1}{3}\pi r^2 h$ 4. $4\pi r^3/3$